The
Final Frontier

by

MICHAEL JOSEPH

LITTLE

ISBN: 1470142015
ISBN-13: 978-1470142018

THE WINDOW TO SPACE IS CLOSING

As energy becomes more expensive food becomes more scarce,
and people continue to multiply

The step into space gets harder everyday

at a certain point
it will get far too expensive for man to make any real foothold
in space or on the Moon

Our destiny is to reach the stars
and it is time

This is our time

the most exciting time of mankind's history is fast approaching
our expansion into space

Let us join together and take this step

CONTENTS

ACKNOWLEDGMENTS

I thank Star Trek
for opening our eyes as to what is possible

1
AMERICAN

We all know it well, life is tough in America today, The boom of the 90's is long gone. I live in Houston and every day on my way to and from work I see many homeless people living outside. It does seem as if life is a little bit tougher than it used to be. After the Vietnam war a lot of Vets were homeless, but that was just a bunch of veterans wondering the country. Nowadays its not just men who can mostly take care of themselves, we see a lot of families and old men and women added in with the Vets from the middle eastern wars. Lots of old men and women.

Americana is LEAN, it has lost ALL its fat. So many people are out of work that don't even get benefits anymore. So many jobs lost as America is just too expensive a place to make almost anything. This is a huge cost to Americans all across the country. We have lost our entire manufacturing base to cheap overseas labor.

Its not the rich people who are suffering, they might have to only buy one mansion in Hawaii instead of two. No the burden of this recession (that was largely caused by rich bankers and speculators) is falling on the weakest of society and those least able to take care of themselves, these are the Americans that are carrying the heaviest burden of the depression. The old, the weak, the sick and the jobless and the poor. Do all of these deserve their lot in life? Do they deserve to be looking through trashcans for food? How many homeless Americans have starved or froze to death. Did all of these dregs of society made strategic decisions resulting in their current position? Did they plan to be poor or homeless?

As Americans, I think we can do better than force old people who have no money to eat out of trashcans. Its not just old people, they found a naked 12 year old girl in a California eating out of a trashcan, sleeping in a BMW. It just makes no sense and the whole thing is just wrong, wrong, wrong. The young, sick, and handicapped are the first to feel the pain and then the old.

Its not the rich bankers who suffer, they may have to cut back to $1250 a bottle wine.

It just seems wrong.

I am a Navy guy and18 of 20 years in the Navy I lived overseas. I found that even though people in other lands look different and speak very differently from us, they tend to think the same way that you or I think. The key is respect, as long as you respect other people they tend to respect you back. Even if you can't talk very well, most everyone does like to drink. Just respect others and they will generally respect you back.

Coming back to the US, I was shocked how little respect people show to each other, a lot of people are just rude. People will stand right next to you are having long loud conversations of sometimes personal nature and it is as if you are not there. Some do have manners, but it is rare. If they have them, they rarely show them.

My family came to America in search of religious freedom and freedom from economic oppression. In our previous land our religion was not the officially sanctioned religion and to believe otherwise was punishable by death.

My uncle has a bible in a safe deposit box, it is very old and has burned edges. It is a Little family bible from hundreds of years ago. This is a significant item of history. The reason this bible has burnt edges is because it was baked inside loaves of bread. The hot oven was the only place the searchers would not search. It was the only place they could hide their bible. It was an outlawed religion and for that bible to be found was death for the family, by order of the king. The bible was burnt because it took my family a long time to get back to the home when they were detained. It was their most prized possession, so valuable they were willing to die for it. They would not just forget. They had to of been detained.

My family has never been rich, and fame has avoided us. A census in 1848 in Georgia shows my great-great-grandfather as a boy of nine. The family declared $3,000 as the worth of the place. A place laid to waste by the war that followed. They then moved to Texas.

In the last hundred seventy years my family has served over 80 years in one uniform (most was in the Navy) or other. We have fought English, Yankees, Germans twice, Japanese, Vietnamese; we have likely fought each other more than the enemy. We have sailed millions of miles and marched a few more. I have spent 20 years in the military and would of served more if I could of. It wasn't for the money, as my pay tripled when I retired from the Navy 14 years ago. I just went right to work in IT. I do love this country and rightly salute every flag I go near.

My job has taken me across the country, to almost every state. It is difficult for me to report, but I must say that America has been in a economic decline for a long time. I have seen so many empty building where once were thriving businesses all across the country.

Its sad because I see these empty buildings as someones broken dream. So many broken dreams. So many lost jobs.

In the last 70 years our country has gone from the top of the economic heap to the bottom of it.

It does not have to be this way.

What the hell happened?

Our corporations have grown so large that they have moved from being country wide companies to global companies and have became very rich and powerful along the way, their influence reaches into governments all around the globe. Then to make their investors even richer they exported most of the good jobs overseas. Over time many stockholders and owners of corporations ran for office. Some won, as they did have more money to run more effective campaigns than their neighbors; some like Hilary sat on Walmarts board and then went on to became the head of the U. S. State Department.

Money or power is not a bad thing in itself.

Bad things did occur because over time these individuals who we elected have made decision after decision in favor of big corporations. In many cases these decisions have resulted in lessened tariffs and jobs moving overseas. Free trade agreements eroded away our ability to get and keep businesses that pay a good wage. It resulted in the net loss of many jobs as corporations gradually moved manufacturing jobs from the highest paid countries to the poorest, they are now exploiting the cheapest and lest protected labor on the planet to make their goods.

It seems as if a global trade war has occurred, that class battles have occurred both globally and locally. Business won and the citizens have been left with paying the bulk of the price of the war. It is also clear that the rich have won and the poor lost.

It has indeed come down to the **have's** holding positions of authority. **Have not's** don't hold many positions of government and when they do, they often side with the rich because that's where the money's at. Money and power corrupt the soul. These well to do people with millions of dollars in the bank are making all the decisions and deciding what kind of scraps they are going to give the **have not's** (citizens). Are these politicians (who remain in office solely by the money that corporations give to them) going to make decisions to benefit the companies stockholders, or the countries taxpayers?

A QUESTION?

Is good for Business,

Always good for America?

These companies give millions to those who see things the way they want them to see things. Its not even illegal. You like the way a politician is doing their job, GIVE THEM MONEY.
Dishonest bastards will follow you forever.
The only thing you can do with a egg sucking dog, is shoot them.

It is becoming class warfare.

The 99 percent don't get it either, they know something is wrong but they don't know how to fix it. Its certainly not going to be a riot or confrontation.

Organization is the key. We can take it back.

One man – One vote

We can take it back but we have to do it vote by vote. If 51% of those who bother to vote, vote one way, we can easily take our government back.

Recently our government tried to encourage the manufacture of American batteries. Then a flood of cheap batteries from China saturated the market. These American battery companies are now struggling to compete with near slave labor wages in China.

It is a economic war and we are losing the battle. The war is still far from over and it could still go either way.

Walmart has wounded American so badly that it is actually hurting their bottom line. So many jobs have left that people can't spend as much as they used too. Walmart forces suppliers to sell so cheaply that they must use cheap foreign labor. Walmart's policies have killed thousands of companies and drove American into near poverty.

Walmart is not alone. Apple is the worlds most profitable company and has 400 billion in the bank. But they use near slave labor to make this money. They have made enough I phones to stretch halfway around the world; yet in the factory that makes them, conditions are so bad that they have to put up nets around the building to keep people from jumping off and killing themselves, because working conditions are so horrid. China has also been known to employ slave labor, their prisons are sometimes work prisons (they could be making your babies toys). China has long employed economic slavery.

We ain't going to **<u>EVER</u>** be able to compete with crap like that.

Not unless we have tariffs.

OK back to why America crashed so hard.

Our politicians have embraced the concept that what is good for business is good for America. And this is not always true.

Is America stockholders or citizens?

Or let me rephrase that
Is what is good for stockholders,
Always good for citizens?

I don't think so.

Our economy is in a shambles. We are recovering from a global meltdown just 4 years ago during the last years of GW Bush. We have a national emergency on our hands and our Congress in Washington DC can't even pass a budget. 99 percent of the focus of almost every one of our elected officials is simply to be reelected at the end of their current term of office. They have absolutely no concern as to the long term well being of our nations citizens. They believe citizens are simply a crop to be bled as white as possible, that its actually OK to take all of the blood of our citizens, their children, and their children.

Sloppy Thirds

The needs of the people come a far third, right after the desires of lobbyist's and corporations. The needs of the people never even get considered, to do so would be political suicide and in the eyes of our elected officials, its just plain STUPID to listen to citizens, they don't know anything.

Shit most citizens can't even keep a job.

How best to get reelected? Get the favor of big corporations that are able to give you millions of dollars than you can use to run for reelection, they might even run the ads for you. These biased individuals then side with big business on any bill that comes up. They might even write bills that favor business. It's all about money and politicians side with business almost every time.
It's all about the bottom line and dividends.

After 60 years of this crap, we have little left in the way of factories. They have all left. We have been reduced to a third world herd of cash cows and we are being sold out by our own Congress.

The American public is now a prostitute that United States Congress is letting China and India have their way with. Along with any other country holding a fistful of money. There is a line out the door.

Does it really make that much sense to have a term limit of a single term of office in any one seat? If you've done everything else, Run for President.

Shit, any idiot can run for President, just be born an American.

It is January 2012 and the nation is heading into a presidential election year. The complete focus of the current election is not the state of our country, an economy in tatters, or a war in Afghanistan, or a very hostile Iran threatening war.

No, not at all, the focus of the election is that the Republicans hate having a black president so badly that they will not cooperate with him on anything and by the way they would elect the devil if that was the only way to get a white face back into the white house. Even a Mormon is tolerable to a black. Perhaps they will then legalize bigamy, then we are going to see some real trouble. One wife is hard enough, two would be murder.

Currently anything that might even remotely be beneficial to our country is up for arguments, debates, and studies taking years; NOTHING GETS DONE. They can barely pass a budget.

Hell, congress is so inept

I bet they have to pay someone to wipe their own asses.

Just plain losers, I rate car salesmen above politicians any day.

A PLATFORM OF CHANGE

Congress needs to realize that American citizen's elected Obama in a landslide because we desired change above all things.

We want to get away from the old way of doing business.

**We are tired of business dictating policy in the
UNITED STATES OF AMERICA**

DAMN IT

THE PEOPLE WILL BE HEARD

We have witnessed a revolt in Congress
Against the legally elected President

The election of Obama has given the largely white or rich Republican party the excuse to just do nothing, to not pass a single bill that the President supports. Completely race and politically motivated. Every single one of these racist self serving Republications needs to be gone from Congress (they really are DONE here).

OVERALL

President Obama has done a terrific job and has gotten us out of one war and has been great for the economy. Remember that the economy failed in George Bushes second term. The economy of the entire country was laying in waste when Obama took the reins three years ago and it has gotten a lot better. People are in stores and shores are open, big banks are making money – not failing. Headlines are generally positive.

President Obama has done this in spite of a TOTAL LACK OF COOPERATION from the Republican Congress. Imagine what he could of done if Congress had worked with the President, instead he had to drag these screaming idiots along for the ride. Can you imagine?

Americans like to work hard and do good, and could of done very well if it had gone in the way the President asked. This was not to be as Congress decided to not do anything the Democrats wanted. So for three years American Democratic and Republican parties have been locked in a political STARE DOWN with each other.

The entire opposition party (the Republicans) have let their emotions rule their voting in many important decisions. They have let their hatred of Obama get out of control and it has been very detrimental to the very health of this country. They put the political desires of the party and the color of their own skin far above the needs of the country.

The Republican Party has revealed also itself to be completely inept and incapable of governing this fine country; they are totally prejudiced against anyone who is different than themselves. They seem to have the herd instinct of Piranha who will eat the weak, except the Republicans eat their own young.

The Democratic Party has also shown itself to favor business and or organizations over the needs of the people. I think they have put equal numbers of republicans as democrats in jail for bribery and fraud.

OUR ELECTED OFFICIALS IN CONGRESS, ARE ALL A COMPLETE DISCRACE TO AMERICA

Beside bitch, moan, and complain; the one thing our politicians can do that takes years. is to go to WAR. So at at the very best our politicians are inept, at worst they are criminally insane.

Any war is completely insane

In WWII we faced Hitler, Tito and Mussolini, in Korea we faced Mao, **in Iraq we had Bush**, our own worst enemy. Bush could of taken on **the real source of Terrorism** in the world and attacked **Iran**, we could of won. Instead, he turns to Iraq and fifty more years of war. It is my belief that any politician that directs this country to war has failed in his primary job of keeping us out of such things, that he is absolutely worthless and should be fired. In war our diplomatic mission has failed and we have been reduced to bombing an entire country into submission and perhaps killing many civilians along the way (who MAY be purely innocent and no danger to anyone).

I see WAR

as the direct failure of our politicians

to keep us out of it

If (a member of congress)

voted for war (in Iraq) they have to go

Now!

War is sometimes justified. Should you or an friendly country be attacked, a hard military reaction is indeed justified. I believe in a **sharp sword for our military**. We should be able to kick anyone's ass, but I also firmly believe in never using force first.

We should not throw the first blow.

I am for a mighty military and quick response. It is certainly OK to go in and kick someones ass; gotta use up that excess ammunition and bombs somewhere, as they do have a shelf life. I do exempt terrorists of any country, it is open season on terrorist. If they are American terrorists, get a warrant and tape it to the hellfire missile you fire from the drone. **But do get a warrant**.

A war for the right reason is perfectly OK, but let the military call the shots, no political calls, AND then go in and and FINISH the opponent off as quickly and possible , THEN get out.

No More Nation Building

America has been at war 72 of the last 105 years. Between WWI and WWII America was fighting an insurrection in the Philippines. If we smash a countries government and see the President hanged we have no obligation to rebuilt the country. We certainly don't have to rebuild it in our own image.

Democracy is great, but people will in the end choose their own destiny. We have no say in this matter and foolish to think we can change destiny.

This is as it should be
Go in hard, Kick ass, and Leave,
Fuck'em (PARDON my FRENCH)

It might sound brutal, but if you didn't really hate them in the first place - **don't go to war.** When we have tried to change history by assassination or war we have always screwed up. Vietnam and Iraq are perfect examples (we assassinated one president, or was it two, and hunted down and hung the other), together these were wars lasted over 25 years and left tens of thousands dead. Many is the pride of the family that came home from a dirty war in a box; fathers, sons, daughters and mothers. If they came home.
How many tears.

To me, the death of just one American service man or women for the wrong reason is way too much cost.

How much American blood

is <u>America's freedom</u> worth?
<u>All of it.</u>

How much American blood

is <u>Iraq's freedom</u> worth?

<u>Not one frigin drop!!</u>

You take freedom
You don't give it away
You can't give it as a gift

I say smash the head of the snake and leave. Use air power and destroy only enough of the offending nations infrastructure to complete the job at hand. If another threat arises, we can always go back. One can try to economically revive a country, but stupid to try to do so where the only thing the countries people hate more than each other, is the USA.

After being married three times. I have learned that one should not try to change your partner. You can't. People will change when they get good and ready and probably never at all. I still try a bit.

YOU CAN'T MAKE ANYONE
DO ANYTHING

So it is with these Muslim countries, but one can affect change in many ways without going to to war. An example is how our very way of life is like a transfusion of new blood into the hearts of many of the old countries of the world such as India and China. These are places Genghis Khan ruled over 700 years ago. History and tradition going hand in hand with our modern way of life. Who would ever of thought that these diverse economies of the world would ever be so intertwined with ours?

Trade with them,
don't go to war with them.

We can change other countries, but the cheapest way to get real change done is by economic means.

Usually it is just stupid to resort to war. It's simply emotions getting out of hand. We should ALSO never support or cooperate with another country that hurts its own citizens. It always comes back to bite us on the ass (Brazil, Iraq, Iran and Vietnam)

BUSH SIMPLY HATED SADDAM

1. The legally elected President of the United States overthrew a legal government

2. He hunted down that countries president, actually found him in a hole in the ground, we dug him out of a bunker with a hidden entrance that was only a few inches across.

3. Saw him hung on gallows that the American taxpayer paid for

We had a madman in the White House, this was clearly a personal matter, not the sort of thing we should of gone to war for. For Bush to go to war because he simply don't like SADDAM is CRIMINALLY INSANE.

I say Prosecute Bush for the premeditated murder of Saddam Hussein

This was clearly premeditated first degree murder,

for Gods sake, at least send him a **BILL** for the cost of the war he wanted so much that he was willing to lie.

THE HUMAN COST OF WAR
IN IRAQ

Is Democracy In Iraq worth even 1 drop of American blood?
No. Sending even 1 serviceman to his or her death is completely
unfathomable for such a stupid reason, yet we lost over 3,531 proud
American Marines, Soldiers, Airman and Sailors who died in combat
in Iraq and tens of thousands more who will carry forever the
physical scars from this conflict. Over 300,000 brain injuries.
Thousands of family's torn apart, many spouses have also died,
as sometimes warriors brought the horrors of war back with them,
in their heads.

The casualties of war are not only in the combat zone.

The echoes of this war resound on our streets as police have to
deal with many who have the demons of war still running through
their brains. Just yesterday six policemen were gunned down by a
single veteran. One died. Who's to blame for this blood?
Bush mostly. But really all of us. We could of stopped it.

The blood of these patriots, their loved ones, and the many
innocents who died along the way are on the hands of anyone who

voted in support of the war in Iraq. My first instinct when I heard that Bush had invaded Iraq was anger at Bush, I knew how stupid it was.

War my friend is not a game to play lightly. In 1945 while we were assaulting Okinawa Japan, in over 100 days of battle we used over 7,000,000 6 inch rounds or better of shells. We didn't even bother to count smaller stuff. This was on about a quarter of a fairly small island. The Okinawan s called it the Typhoon of Steel. We bombed a big city (Naha) into nothing except craters and destruction. Almost 200,000 civilians died. It is my thought that wars of any magnitude should generally be avoided.

Companies like wars, they make money.

We need NEW blood
in United States Congress

GET RID OF THEM
ALL OF THEM

When you go to the polls, remember to NOT vote for anyone who supported going to war in Iraq (Democrats and Republicans are both guilty, get rid of them all).

Oh my god. We have elected an entire generation of Benedict Arnold's to Congress. Traitors all. If Andrew Jackson or John Adams were alive this day they would take a gun and forcibly retire the whole worthless lot. All they want to do is make money and get reelected. No desire to do something for America. It's all for them, their friends, their benefactors; it's all for money. No heart, no soul. What complete losers. Traitors all.

Will the Lord please have mercy on America and set a term limit for any office to 1 or 2 terms. So the worst these worthless leaches can do is bleed us for only 2 terms.
Not for their entire frigging life.

I guess we know how Rome fell, it wasn't forces from without – no their own Senate got too greedy and took it down from within. They fell and everything came down with them. The Gauls and others were just taking advantage of a weakened Rome.   A Rome where the politicians became so corrupted that nothing could ever get done. They could not figure out how to pay for an army to fight off the Gauls, much like U. S. Congress – full of idiots.
China and India and anywhere labor is cheap is our Gauls.

I hate to agree with Karl Marx on anything, but it is clear that capitalism is not always good for the masses. Not like it is in today's world when capitalists have taken over. Its sad, but frequently decisions are make in governments that favor business over objections of its own citizens. The needs and desires of the people are not heeded at all.   I do not believe its related to party, politicians of all creeds and colors have sold out to business and crime. Some just put out occasionally, while the greedy never stop sucking that corporate money spigot. They do it all the time.
Figure which who or which party does more. I can't.

Perhaps it will take blood in the streets, like what happened a hundred years ago. Before labor laws cleaned up a lot of abuse by business, it took riots to stop the orchestrated violence by business towards workers. Many were killed, by both sides.

Its hard to conceive but it only took 30 years of mostly incompetent and greedy politicians to drive this great country into the ground. At the time free trade sounded nice and true to the spirit of the strong American free enterprise system that we all love so dear. The working man stood up back then and said no.
Hell – they actually DID stand up and say hell no, big protests.
Our honorably elected officials then got into bed with corporations and let the stockholders have their way with the American citizen. They told them that they knew what was best for them. Because what is good for business is always good for the citizens – right? Wrong. Wrong. Wrong.

You don't agree? How many stockholders are their in America? That own 90 - 99 % of the wealth? What kind of percentage here? I bet no more than 5 % of the population is controlling 99 % of the wealth. This is the percentage of Americans benefiting from free trade. While the other 95% have lost a lot of REALLY GOOD jobs and the American Public have simply become a cash cow to be milked dry. We are pretty dry. We are just Walmart shoppers.

Walmart Mandates Pricing

Walmart the worlds biggest retailer mandates prices to its suppliers. In many cases these mandated prices are much less than can be achieved using American labor. They are squeezing these suppliers who make very little (still they get rich). If you want to do business you will likely manufacture overseas, this forced American companies to produce goods overseas.

In far too many cases corporations have first moved just a few manufacturing jobs overseas – then followed with all of them. Business will not manufacture stuff in the US under those conditions. They have no incentive to stay here in America.

When we leveled the playing ground with both India and China, vast sources of cheap labor became available and this was a boom to the stock holders of corporations exploiting them. Then Walmart took off and they demanded that most of the goods in their store be produced in China. Business boomed, but then the jobs left the good USA because we are not competitive in the labor market. Its hard to complete against cheaply paid labor in China and India.

Tariffs should be raised to equal the playing field but this is not even an possibility where past and present board members on the

Walmart board and many other big companies sit in high positions of government. While the desires of business gets heard and heeded, the voice's and cry's of the people are not even considered.

If it does not add to the bottom line, it is a threat to business and America.

War is Looming

When oil supplies are threatened, war is near. So far pressure from business and stockholders have lead us into two wars. Another war is looming as the USA and Iran square off in the Straits of Hormuz.

The last time the Iranians went to war they were sending battalions of 13 year old children against entrenched machine guns. The parents were even happy. If it is not a very quick war, expect it to be very bloody. Under all outcomes expect a bloodbath.
It is definitely a lose lose situation. Definitely a hand off to Israeli. Earn your pay.

Do we have better things to do than send our children to fight another country's children.

Global communications have changed the planet, it enables corporations to use the cheapest labor available on the planet for making any product. In America, not too many years ago manufacturing was in swing, wages in these factories were far better than average (likely a reason they closed so fast after the free trade acts). Entire regions were booming. Before we had blue color workers who were paid good wages in return for a good days work. Manufacturing is great – you are essentially making something out of nothing. This is the best form of capitalization. Both companies and unions got greedy and companies either closed or moved manufacturing operations overseas to third world nations.

The stockholder won – big time.
Cheap labor = high profits.

An offshoot of this type of manufacturing is how vicious these companies can be. Forced labor, prisoners, child labor and economically challenged labor, illegal labor. Many people around the world are indentured servants of the system and will die doing what they are doing right now. This is a misery felt world wide. Few countries are booming in this recession.

The good news, business is booming. Stockholders are doing better than ever.

It does not matter all the manufacturers have left for China and people have lost their high paying jobs and many must find jobs that pay a lot less than they used to make. Our children will certainly make a lot less money than we did.

The Taxpayer lost – big time. He even had to get a lower paying job and a second lower paying job.

Our politicians have made so many concessions to the devil, that there is nothing left.

America is done

We might as well all lay down and die.

<u>Bullshit</u>

(forgive me my French, I am also rude and smelly. Ask any of my wives, former's and present).

Lets step

THE HELL *<u>OUT OF THE BOX</u>*

MICHAEL JOSEPH LITTLE

2

TIME TO MAKE BIG DECISIONS

How to rapidly pull this country out of the slump it is in? Kicking the economy in the ass is the only way. Let's do something we have not done in 70 years, fully mobilize on multiple fronts on a national scale:

NEW GOALS

1. **Bring our military home**, close overseas bases

   Every major American city should have it's own air field, naval station if on the coast, and army base. They would all wear the same uniforms they wear now, but ships, planes, tanks, and personnel would also proudly bear the name of that city; these units would be mostly filled with people from that city or area. Unless deployed these soldiers could generally could go home at night and see their extended family on holidays. Lets build family. We spend a lot of money supporting American troops overseas, a lot of money leaves US shores and does not come back. This is a significant amount of money as these soldiers and organizations on these bases spend gobs of money. I want this money to go back into the hands of Americans.

Unless specifically requested, lets not remove our military members so far from home that they lose contact with their immediate family. Stop destroying the social fabrics of families whose members are in the military. Deployments should be rare. All our branches should be able to rapidly deploy anywhere in the world; go there, do the job and come back home. The Civil War was the first war to really tear our country and families apart. We lost our innocence. We need to get it back. But stay well armed.

2. **Integrate Our Bases into Our Cities** there are existing building on bases all over the country that could be partially converted into high technology centers and business centers with the goal of teaching and utilizing technology for the military and civilians in the city.

I want the military to make a profit. If the Red Chinese Army can make a profit, so can the American Army, Air Force and Navy. For ventures to the moon, have them issue stock like corporations do. As an example I am thinking an **investment in the legendary 7ᵗʰ Calvary is a pretty wise investment**. These organizations could be our front line troops in our venture into space. We can teach veterans and civilians how to be productive into the space age, build schools and factories on the

bases that are integrated with companies and industry.
Interconnected with our drive into space.

3. **Invest in our cities and infrastructure**

that are falling apart with the goal of creating eco friendly
self sustaining cities that use as little gas and oil as possible,
for transportation: walking, biking, subways, trains, and buses
should prevail. I would like to encourage cities that are small
enough that everyone should just walk or bike to work.
Connect cities by train. Same kind of system our grand fathers
used. It worked for them, it can work again for us.
Get small and green. Encourage small towns.

4. **Build a strong technology base** in our own

cities that helps support the next big step, the step into space.
The step into space will create thousands of contracts for various
types of personnel, equipment, and support. Direct these to
companies in our own cities and business. Encourage growth and
innovation of new ideas along the way. Just the flights to the
Moon will require making a bunch of Apollo Saturn 5 rockets,
these will be fun bad boys to have around; or else we have to hire
a bunch of Russian heavy lift rocket launches to get our stuff in
orbit. I say make some Saturn 6's, bigger and badder than anyone

has every seen. Build them in the U. S. and require 100% American stockholders, companies, content, and labor.

5. **Let Money from Business and Investors Lead** the Way back to the moon, we want to make money and energize our economy. HAVE SHARES, BONDS, INVESTORS, BUSINESS, AND INDIVIDUALS pay for it, government picks up the rest. ALL are investors and hopefully well rewarded in the end. Like the colonies of early America, these toeholds on the moon and beyond are investments.

6. **Hazardous Duty** some functions of Space Exploration may be so hazardous that its hard to keep workers and yet so critical they need to be manned. In this case we might see convicts working in space for reduced sentences or be under military control. Or just highly paid workers. Maybe both.

America is not dead, we have just been steered into a dead end by idiot politicians. We need consensus, **let us take the first step to the moon**. Bring all our friends along and fully U. S. owned corporations too. Go back as an investment in the present and future.

LET US GO BACK TO THE MOON AND **MAKE MONEY** DOING IT

This is a world effort. We should work with all who are interested in space.   If we can get a moon base going we can shoot material right up to the different Lagrange points. We just have to cook the soil of the moon to get air and water. This will enable us to colonize the Moon and to supply goods and materials from the surface of the Moon to habitats and manufacturing facilities in orbit around the Earth and Moon.

Its a neat trick, you throw up a ball and it don't come down

Many would say I am crazy and close this book. I have spent a lot of time working with steel and think that steel is the way to go in space. It is very strong and able to withstand the rigors of space.

Mike Has FINALLY LOST IT

1. There is no air on the Moon – Duh.

2. That lack of air means we can use a magnetic gun to propel iron objects off the face of the Moon. We build the gun here on earth in pieces, rocket it up to the Moon and erect it. No air to slow down the projectiles. This is not rocket science, we can build it today.

3. Approximately 20% of meteors are iron. This makes for a lunar surface that is rich in iron. We can pick up a lot of this iron using magnets on remotely controlled semi-intelligent solar powered mining vehicles. We use this iron to make cargo containers out of iron.
Make hollow iron spheres that we can fill with cargo. These we can shoot into space.

4. Imagine shooting a iron sphere full of liquid oxygen, or iron pellets straight up off the face of the moon, and it don't come down. It is a neat trick. It is feasible now. We shoot these cargo filled iron spheres up to Lagrange points.

Business and people can pay, governments foot the rest. We do not have to pay for it all ourselves. Each country that wishes to make a significant foothold in space needs to take a section and run with it. We do need to coordinate standards and activities.

America, Canada, Briton and Australia take the two sides of the Moon, while Japan takes the south pole and Korea at the north. China and India can chime in wherever they want. Plenty of room.

We need a global GPS system around the Moon and a wireless global lunar Ethernet system using many orbiting satellites. We will also have to create a target and recovery system in stationary orbit for cargo coming up from the moon, these cargo spheres would be drifting at just a few miles an hour (as they would be shot with the exact speed required to just leave the gravity of the Moon.

The main benefit for those left on Earth is the huge number of jobs that such an undertaking creates.

In the vernacular of the Navy,

"We are going to need a buttload of stuff".

Its going to take thousands of suppliers working for a long time to provide this stuff. Let us make sure that these contracts go to American business and are manufactured on American soil. We might even identity small towns as the place for these businesses to be, we can encourage the growth of thousands of small towns and create literally MILLIONS OF JOBS. Good jobs, not your typical McDonald hamburger shop job (any job is good, you just grow out of wanting to do some of them).

We will be creating spheres in space. Out of steel banding.

We launch tight rolls of steel banding, from these we form spheres in space that we can make different types of rooms. Engine rooms, control rooms, cargo rooms, bathrooms, green houses and on and on. To make a freighter you might have an engine room, living quarters, greenhouse, control room and lots of cargo rooms.

The spheres we can make in space, but it will take a lot of stuff to outfit these sphere to be the type of rooms we want them to be. In the first few years of space exploration, this stuff will have to be made on the earth and shipped up to space. Lots of contracts, factories and a lot of rockets, thousands. Maybe we can make Saturn 5's again. Some nice BIG American rockets.

We can build 40 foot spheres and connect them together to both support each other and to form much larger structures in space. Imagine a tube shaped structure made from of thousands of these spheres and is a over a mile across and spinning (to provide gravity to inhabitants). For a smaller sized one, say 500 feet across you could enclose the whole tube with tightly meshed steel banding and have a ecosystem full of air and water. It would indeed be a magical place.

We can find iron on the moon. Using Vacuum casting we can cast this iron into small cargo spheres, these we can fill with oxygen, hydrogen, and water and then shoot up (maybe one every second, using a magnetic gun) into the stationary Lagrange points located right above both the near and far side of the moon. We would catch these cargo pods in space and smelt them into steel using huge magnetically controlled solar ovens. We could create a huge white hot glob of thousands of tons of molten steel just floating in space. We could use magnetic taps to guide this hot and positively charged steel into negatively charged casting of any desired size. I think we could cast the inside of a steel spheres, make real iron spheres in space, big ones. Heavy ones. Imagine making a yacht of 50,000 tons or a warship of 200,000 tons. Its all steel that came up from the moon.

3

SOLEVEN

This has been an ongoing dream of mine for 30 years, it becomes ever more solid to me as I ponder what man can do in space. I think of a cosmopolitan city in space named Soleven, with all the hates, loves, and dislikes of any busy sea port.

When you arrive this is a city that you are so excited that you run ashore, every time. But at the end of a port visit it is a place that you want to get away as quickly as you can, you run away from it, it is a place with dark secrets and screaming desires.

On Earth we live in a world where air and water are clearly divided. Water falls to the ground and puddles up.

In space in a sphere with no gravity; water and air are equals, no puddles. Surface tension will gradually bring most goblets together, but this takes awhile. It's almost like living underwater except that you can breath. I envision a central sphere being used to store heat, oxygen and water. Under zero gravity a big pool of water in a sphere acts rather different. Under zero gravity, air and water become mixed; so much so that if they were in equal portions of air and water, fish could actually jump from one goblet of water to another. Shrimp, Crab and lobster would also likely thrive in this environment, they would be the garbageman.

Imagine eating king crab from animals over six foot across. When they die we can use the shells to make a form of concrete, same of our bones when we die.

Waste not ,want not.

Imagine hundreds of spheres, organized into a multitude of shapes. Some hundreds of feet across, some thousands. Each sphere a hive of activity and a virtual explosion of human activity.

A sphere might be anything; cargo hold, hotel room, lounge, living room, house, control room, engine room, greenhouse, bunk room, kitchen, bathroom. Almost any size, anything.

In space it is smart to use a sphere to build things as they are very strong. Connect them directly or with tunnels of braided steel bands, these would be made to flex.

You DONT USE SQUARE ROOMS IN SPACE, IT IS JUST WASTE. Its just a lot smarter to make round things in orbit.

You also put spheres together anyway you want, for a tug boat you might want a couple engine rooms, greenhouse, and a control room. For a home you would want the same as a tug would need and add on a couple greenhouses, a couple rooms and lots of bathrooms.

We could connect a bunch of spheres together in a ring and set it in motion, like a big spinning chocolate donut in space. Made out of hundreds of spheres. A big rolling tube of spheres mounted to each other. Each rotating fast enough that people inside the tube could feel a comfortable gravity. Start with one ring and just add to it.

It would be nice to make a tube of spheres and cap the ends, then layer the outside with steel banding. Create an airtight hull and the you can fill the center with air. For a hospital or school, one might also create an inner core, then put layers around it. Like an onion.

These structures might be sponsored by governments or organizations. Each could have their own laws and government, or they might have no law. They might organize many spheres under one flag. Or they might be completely independent with laws unique to their sphere. Not very many things would be illegal. Assault would be illegal as well as murder and stealing. Punishments would be swift.

Spheres from the Moon can supply water and oxygen to habitats in space. These supplies can be shot directly to these habitats using iron canisters.

In the common areas of Soleven it gets a lot more complicated, as these are the areas where different people buy or sell goods. It would be a free market for those who either live on Soleven or travel to Soleven to buy or sell goods. I would also expect military craft and craft from other habitats in space. People from all over would get mixed together. Earth raised humans would be heavy and strong while those born on the moon would be light, weak, and pale skin.

I bet a chicken raised in zero gravity would be fast as lighting. Those born in space and raised under zero or low gravity would be very different from the earthers or mooners. Those born in space might have long arms and very weak and long muscles, with very light skin, I think they would be very fast also, but break easy. They would be much faster than heavy clumsy earthers.

Factor in all the differences in body and colors of humanity and you get a circus. I once described culture as mans celebration of life. Soleven's central areas would be a vast multicultural celebration of life in all its incarnations. A place where people from all over come together in the voids of space.

When you find your way into a bar in Soleven, all bets are off. You don't know what you are going to see.

Spheres are connected by flexible tunnels woven out of multiple steel bands. Air, water, and heat flow in and out of these tunnels. Each tunnel and sphere has automatic shut offs in case they rupture into space.

For sheer redundancy it would be best for all spheres in a spinning array of hundreds or even thousands to rely on multiple sources for water, food, oxygen, and heat. Would build them around multiple greenhouses and heating spheres, if one source failed, another source could simply take that load up. No central point of failure.

I envision Soleven as a spinning tube with an intricate fabric of interconnected layers of spheres. Arranged like stacks of old tires. But these are steel spheres and tunnels and we can arrange them in such a way that we create a huge magnetic field, this can be energized to maybe protect us from solar storms.

Soleven would start as a spinning wheel but would grow into a tube and it gets much larger and more and more spheres are added to the sides and ends. It would be very economically based, towards the center is cheap and towards the outer of the the tube, it is more expensive due to higher gravity.

4

LET US GO BACK TO THE MOON IN A BIG WAY

The moon race of the 1960's forever changed the world. IN 2014 we can step into space and man will no longer be Earthbound. For better for for worse, mankind will be unleashed into the vastness of the Universe. We will go slow at first, then gradually go faster and faster. In the 60's the simple need to make a a small programmable calculator lead Texas Instruments to create the first computer processors. These small electrical circuits led to the PC and to the interconnected world we see today.

Lets make an investment

Let us go back to the moon and beyond

Lets do it for the United States

We can't do it alone

We need a world effort

I am talking global effort. We have a lot to do and it makes no sense to do it alone. WORK TOGETHER.

We Need:

1. A GPS system around the moon

2. A wireless communication system in orbit around the moon so that every inch of the moons surface would be covered by high bandwidth communications with the earth (with a 5 second lag due to the distance)

3. Development of a remotely controlled solar powered robot vehicle that is programmable and able to mine the Moon's surface, continually covering an assigned area. It would have a long solar heated pipe oven, a screw inside would pull compressed moon dust and soil into the oven and the heat would rise to near 1,000 F, the soil would release gases such as oxygen and hydrogen, as well as water. These gases would be captured (and compressed into liquid state at night)
During the moons day this ROV would roam the surface and continually search for iron material. Using vacuum casting the ROV's should be able to melt and cast the iron pellets that it finds into standard sized cargo containers. These it fills with cargo like air, water, or

metals. The moon soil is rich in these minerals and gases.

4. Magnetic guns get made in stack able pieces – Like heavy Lego's. Or maybe lift the entire gun from Earth and launch it to the moon and use the lander itself as a base for the gun. As ROV's bring back full cargo containers they get directly bought and most of these are rolled down a shaft that ends in the launch position for the gun. No warehouse. It get's shot right into orbit.

5. Need to build multiple large solar electric panel arrays on the moon, to power the magnetic gun. These have to be pretty close to the gun.

6. In space we need to design a Solar Iron Smelter that can run under weightless conditions and in a vacuum. It is possible using a big ring of smaller rings of magnets. Like the collider in France. Only this time we are using iron pellets. A large parabolic reflectors concentrates sunlight onto small areas of the big ring of magnets that are straight and require no controlling magnetic field to make it go straight. Over time the pellets in the flow get so hot that they melt into molten iron (steel). These streams of molten iron are controllable due to their magnetism. This ring of molten iron can be magnetically

tapped in the same fashion. The smelter can never cool, unless emptied first and this would likely be rare.

Iron and Steel

This ferrous metal helps make our world: our cars, our toys, our tools, many are steel. The work horses of our civilization, our ships and trucks are made of iron and its harder brother steel. Our aircraft carriers are a monument towards the things you can do with steel.

Great Memory, Steel and iron remember their shape very well and are adaptable to almost any shape we want it to be. Spheres are rather easy as they are just a bunch of hoops put together, or more likely just be a single ribbon of steel rolled in the shape of a ball.

Steel is an angry metal and is dangerous to work with, since it requires such high temperatures to melt and has sharp edges. But pound for pound it is the easiest to work with and strongest by far of other metals. We can bend it, weld it and cast it.

We can use Waldo's in space to do the work. Intelligent remotely controlled robots that we can wear, they have a near and far side. Dangerous work is done using the far side. It can also be scaled a lot bigger. We can make remotely controlled iron or steel giants. We need these Waldo's.

The Magnetic nature of molten iron makes it perfect to work with using magnetic force. We can control the movement of a molten stream of iron or steel. It is also possible to TAP this molten iron with other magnetic tools.

It is preferable to have a stationary molten glob of steel in space, one that is ready to cast into whatever you want it to be. It could be a white hot globe of metal and be used as a huge light bulb to illuminate the inside surface of a rotating wheel. Say 200,000 tons of molten steel.

MICHAEL JOSEPH LITTLE

5

AN AMERICAN MOON BASE
ON THE FAR SIDE OF THE MOON

I propose the United States, Canada and Mexico create a Moon base on the FAR of the Moon (Dark Side Of The Moon)

We will need: **A Trading Post** A maned or unmanned trading post on the moon that can sell, rent, and maintain remotely operated mining vehicles (ROV's) to operators from Earth. These would go out and harvest iron pellets from the surface,   these would be melted and cast into 1 meter iron spheres that are then used as cargo containers. During the day these ROV's move around and continually pick up lunar soil and run this through a long solar oven, this releases gases such as oxygen and hydrogen, these would be compressed. During the cold night these gases would be reduced to a liquid gas and packed into cargo containers (along with excess Iron and lunar concrete) and sold to the trading post.

The Trading Post buy's trade goods from the ROV's that occasionally visit the trading post and sell containers full of oxygen, water, gases, iron and other metals and gases to the trading post.

A **Manned Service Facility** is required for servicing, maintenance, and repair of the ROV's. I expect these guys to get paid pretty good.

A **Magnetic Gun or Two** is required as we should be able to almost immediately shoot iron cargo containers up and into L2. We don't have to invent much as we have the ability right now to throw iron objects from the surface of the moon up to L1 and L2 from the surface of the Moon. A Magnetic gun can do it. This material could very well be used to support industry in space.

L2

is far above the dead center point of the FAR side

(Dark Side of the Moon) DSOM

L1

is far above the dead center point of the

NEAR Side of the Moon

Far above our proposed Moon base on the

Dark Side Of the Moon

I propose the creation of a space based

shipyard in space at L2

to make **ships & habitats & greenhouses**

in the weightlessness of space.

Gateway to

Solar
System

L2

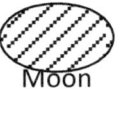

Moon

L1

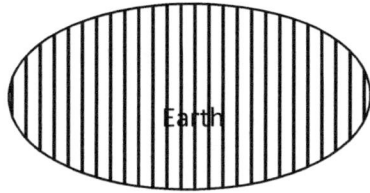

Earth

Graphic 1. Where L1 and L2 are located

Looking at Graphic 1, L2 is the point directly away from Earth on the FAR side of the Moon, far above the surface of the moon, it is called the Gateway to the Solar System. So called because it is so easy to move from L2 into interstellar space, away from the gravitational influence of both the Earth and the Moon.

This shipyard would have a recovery target right above the magnetic gun on the surface of the moon. This recovery system would have a maneuvering system that would continually center itself on the stream of cargo containers being shot upward from the moon. **The gun and recovery area in space are stationary to each other.**

METEORS HIT THE MOON HARD

One way to get resources such as water, air liquid oxygen, and liquid hydrogen into space is to use iron cargo containers. We can use a magnetic gun to quickly accelerate these objects faster than 10,000 feet per second. The cargo does not touch the gun. The gun is shooting an a vacuum and you don't even get a sonic boom as the iron sphere breaks the sound barrier. Its just gone. The bullet does not even touch the barrel, no wear.

Iron is the Key
From Iron we can make Steel

The key to industrializing space is iron, this is critical to making structures that are able to last a long time. It is the only thing that is tough enough to stand up to the rigors of space. About 20% of the meteors that hit the moon are nickle iron. These hit the moon going at least 8,000 feet per second (this is escape velocity for the moon).

Since the moon has no atmosphere (and no Terminal velocity) meteors falling to the lunar surface will increase up to this speed or maybe even faster (depends on how fast they were originally moving).

As meteors hit the lunar surface at 2 miles a second they generate a huge amount of friction and heat, an iron meteor will partially melt on impact and a deadly splash results, some of this molten iron is thrown outwards as goblets of iron, these fly outwards about the width of the impact crater, cooling and hitting the ground as showers of iron pellets. This looks exactly like dropping a rock into a pool of water from a great height, you get a nice splash. This type of impact has been recreated in a vacuum with high speed cameras. This effect does happen. WE CAN MINE THIS IRON.

We can mine likely areas first for optimum yields and we should be able to isolate the regions with the greatest magnetism and use magnetic mining arms on remotely controlled robots to pick up these iron pellets. On the surface of the moon these pellets can be cast into iron spheres that can be used as cargo containers. These can be the bullets in a magnetic gun able to fire them into orbit.

A magnetically controlled Furnace

In orbit, our shipyard would take iron cargo containers and iron pellets, and melt them using a solar steel furnace. This solar furnace would be circled by rings of magnets that guide the molten iron / steel in a slow moving continuous stream, a river of slow moving molten metal that just goes round and round. This molten river could be magnetically taped and used to cast large iron spheres and press out steel banding.

Taps can direct this molten metal to casting vessels. Some as big as spheres. Imagine casting a space ship of 50,000 tons of steel. Cast it out of steel and do it as a whole.

Steam generates lots of force, we can use steam or rocket engines. Steam is much easier to control and you don't need liquid oxygen or hydrogen. Both very dangerous to work with. I say we use steam to travel in near earth space. Can't be in a great hurry.

There is really nothing better than steel to build from, especially in the harsh vacuum of space. You can weld it and it is extremely strong, even one layer of stainless banding is good for holding in an atmosphere. You could even put a hole in a sphere using a magnetic gun and the sphere would likely not rupture. You could fill the hole and it would be structurally sound again.

Depending on end use, spheres could be made from everything: iron, steel, carbon fiber, concrete, steel banding, and ceramics. These spheres could be utilized for multiple functions and used to made almost anything, such as: control rooms, engine rooms, water and gas tanks, living quarters, hydroponic tanks, even hotel rooms, or homes. They can be big, likely much bigger than 1000 meters across. 10 meter would be a standard.

Spheres could be finished in multiple ways such as making a control room or engine room. The options are endless and many of these systems would be lifted up from earth.

For Hotel lobbies and rooms I am sure that windows to the stars would be very popular. We could not really face the sun because its very dangerous, but we can use reflections of it. Views towards the distant stars would be nice, the north star would be very popular to sight on. Most of these views would be digital as, living in free fall is such a drag that most of your life you would spend in a rotating wheel of spheres. Real star are only really viewable when you are in free fall.

Most of the time you would be spinning as you live in a centrifuge to simulate gravity you are unable to comfortably see the stars, as they would all be spinning madly.

MICHAEL JOSEPH LITTLE

6

THE NEAR SIDE OF THE MOON

I propose the British, Europeans, and Indians operate a Moon base on the NEAR side of the Moon Trading Post In the same manner as the Americans. The idea is to make money.

The Point of Sale trading post would buy trade goods from the ROV's that return to the trading post and sell oxygen, water, gases, iron and other metals and gases to the trading post.

A manned Service Facility, a maintenance and repair facility would service the ROV's.

A Magnetic Gun A magnetic gun would be able to immediately shoot these iron cargo containers into L1. L1 is far above the dead center point of the NEAR Side of the Moon

Managing L1 They would be using L1 as a transit and refuel station for vehicles to and from Earth. L1 is a bit more unstable compared to L2 which is very stable. Still very do'able .

In addition I propose the foundation of a Sanctuary, Academy, and a Hospital on both the NEAR and FAR sides of the Moon:

Sanctuary Many men and women will be needed to take the huge step of travel into space, and babies will surely be born, these children would be the first members of the human race born off planet and they deserve a haven, a chance for life.

The **Academy** would be a cradle for those born away from Earth. Mothers and Fathers would be free to stay and the family would be given a home that would forever be owned by the bloodline. If they leave they are welcome back anytime they want, it's not an offer, it's a right. Non transferable.

A Hospital would provide care. We are going to have accidents.

Anyone would be welcome, no one will ever be turned away, but all must work and earn their keep (if they can). In space one must either make or earn the very air that your breathe. But on the Moon, none would be cast out. In the event of a major crime, the laws of society might decide to revoke sanctuary, but only after a full vote is taken. An option might be self expulsion or self imprisonment. Or take a dangerous job. But if you are able to work, but choose not to, you would not earn any credit towards off world goods.

The Hospital, Sanctuary and Academy would each get 1% of any cargo leaving the surface of the moon. The riches of the moon would ensure that these facilities would never have to turn anyone away. It would become a home, a school, a tradition for all to enjoy.

Sanctuary is one of MANDKINDS INSURANCE POLICYS, that man can survive any Earthly disaster.

Today if volcanoes Yellowstone and Toba both decided to erupt, mankind would likely disappear in just a year or two. When the sun goes out for longer than plants can survive is a time we all could die.

MICHAEL JOSEPH LITTLE

7

SETTLEMENTS AT THE NORTH AND SOUTH POLES OF THE MOON

Settlements at the north and south poles of the Moon to look for water. Let us drill and see what we find. Let the Japanese and the South Koreans have the South pole, and let the Chinese & the North Koreans can have the North.

Like everyone else on the moon, they would use ROV's as mining robots and can shoot cargo up to either L1 or L2 in the same manner as the Americans and Europeans British.

A Magnetic Gun to shoot cargo up to L1 or L2 from the Moon's pole's would be laying down and shooting almost parallel to the the ground. Not a problem, no air to contend with.

Finding lots of water and shooting this water to L1 and L2 is critical to space colonization as water is perhaps the most precious resource of the moon and this can be shot directly into space.

Water can be used as a propellent. Under heat and pressure water is a very effective propellent. A ship of spheres using water as a propellent could definitely go at least 1000 miles per hour (not much more). If we have enough water in space for everything else, water can definitely be used for propulsion. At least as long as we are not in a real hurry. It's much safer than mixing liquid oxygen and hydrogen together in a rocket engine.

8

HIGH GROUND

Does the United States of America care if someone else develops a Moon base or space station before we do?

Hell yes !

Would you feel comfortable with the North Koreans and Iranians getting together and creating a space station or moon base? The tactical implications of a hostile force in or near earth orbit and or located on the moon are HUGE.

Imagine Iran or North Korea having a moon base and they just start shooting 500 pound iron balls filled with shot (like big shotgun shells), at people or things they don't like. Even with one gun, you use time on target shooting (varying speed and targets) and you shoot in such a way that all your targets get hit at the same instant. Its very easy to make a gun out of magnets capable of sending iron objects from the surface of the moon to the distant earth. You can hit a fairly small target. As long as it is a stationary target.

OK time for a deep breath – now turn the page.

Pull back to America
centralize our mission and
set our goals for the next 20 years

We have no long range goals. If it takes longer than 2 years to build, it just does not get done.

Our own traitorous government is allowing the U. S. to fall behind in many key scientific and space areas. We have no rockets or plans to build any real soon. Its as if we are being lead by donkeys following carrots on a string. One big one is the race into space and on to the moon, we are not in it. The Chinese will be the ones that colonize the moon. This will not be too many years away.

What is the implications of a Red Chinese Base on the Moon?

ALL BAD

We were victorious in the latest wars in the middle east, in Iraq and Afghanistan. These were against a nearly stone age people who had been slowly derogating in culture and society for hundreds of years, becoming ever more savage with each passing year. We just helped them along. We won because we have the capability to drop bombs anywhere we want, anytime we wanted too. Ground troops could simply laser a target and soon it would be destroyed.

In a war zone it is really convenient to have a couple of B52's circling an area carrying hundreds of laser guided 1000 pound bombs.

They just dominate the battlefield, they can't even see the ground but they are the thunder gods above. It is impossible to see them they are flying so high.

In a Like Manner who Controls Weapons in Space or on the Moon Controls Earth. I don't know about you, but my Chinese, Korean, AND Arabic sucks. I am sucking here.

Who controls the moon

Controls Earth

Many groups would like to control the world. Some of them can make Hitler look like a school boy. Some would destroy it all.

In the same manner of a B52 dominating a battle field, control of the Moon or having military platforms in space gives the owner the ability to target a specific target in any city on Earth with long range bombardment by rail gun's or arrays of magnets mounted either on the Moon or in space. These would be rapid firing high speed systems that would use Time on Target shooting, where they can precisely vary the bullets velocity and can shoot at multiple targets that all get hit at the same time.

They could launch nuclear weapons from the moon.

Unopposed control of a electromagnetic gun on the Moon gives the owner the ability to quickly hit any target on Earth that looks like it could be getting ready to launch a moon rocket. If a manic got control of the system, he could easily target rocket pads, smoke stacks, bridges, and airports and quickly reduce mankind's level of technology to the stone age or very close to it.

Control of a electromagnetic gun on the Moon or in a weightless position at L1 & L2 gives the owner the ability to try to protect the Earth against many possible DOOMSDAY asteroids. We could use a quick loading system and be able to fire nearly continually, (shooting iron or steel containers at 10,000 feet per second and 60 rounds a minute), we could really hammer any possible threat to the

Earth and our habitats on the Moon and in space. Might help. It could also hammer any unfriendly target in space or on Earth.

What is going to happen when the Red Chinese, North Koreans, or Iran get weapons on the moon?

BAD things, that's for sure.

A hostile adversary on the moon could quickly hit any target on Earth that looks like it could be getting ready to launch a moon rocket.

MICHAEL JOSEPH LITTLE

9

LIFE IN SPACE

On the surface of the moon let us look for iron using remote controlled vehicles and magnets. These would be solar powered intelligent vehicles, with full communications to operators on Earth. They will roam preassigned areas and continually pick up soil that is baked in solar ovens to release captive gases – these will be saved. Arrays of magnets will pick up iron pellets, these will be melted and cast into cargo containers that can be filled with gases and metals and be able to be shot off the surface with a a magnetic gun.

Lunar iron would be used to create 1 meter round empty iron tanks (they can be vacuum cast), this would be the standard lunar cargo container size, able to be filled with water, liquid oxygen, or hydrogen, or filled with iron pellets to be further refined further in the solar furnace at L2 into steel.

As ROV's rove the surface during the hot 14 day lunar DAY: Lunar soil would be placed into a compacting bin to be compressed and drawn into a corkscrew heater in a long solar oven, it is already 300 Celsius on the surface of the moon, getting to 800 Celsius is not that hard. This heated material releases oxygen, hydrogen, (water) and other gases – all is captured and compressed.

During the long cold 14 day NIGHT: compressed gases could easily be reduced to a liquid, these liquid gases can then be packed into iron cargo containers and fired up to L1 on the near side, and L2 on the far side of the Moon. I can imagine that a meter sized container of liquid oxygen or water would be quite valuable in space. But would gradually get cheaper as more and more of it get shipped into space. It is here on the Moon that the first lunar millionaires will be made. Owners of the remotely controlled solar powered mining vehicles will reap huge continuing dividends.

One meter iron spheres would be launched upward using magnetic guns. The spheres never even touch the magnetic gun (no wear).

The Moons L2, if you stand right below this point and throw a ball up at 5,400 miles an hour (about 3 times as fast as an M16 bullet), that ball escapes the Moons gravity field and starts floating out into the Solar System. This is the place in the solar system called the gateway to the solar system because it takes so little energy to move out into the galactic plane.

Benefits of L2

1. You are not orbiting the moon If you stand on the surface of the FAR Side of the Moon and throw a cargo container upward at the escape velocity of the Moon (plus 1 mile per hour extra), this cargo container will escape the moons gravitational field and drift into L2 where it can be recovered.

2. You remain forever still in relationship to the Moon and the Earth. The center of the dark side of the Moon is always directly below you. In relationship with the Moon and Earth you remain completely stationary.

2. The center of the dark side of the moon is very VALUABLE REAL ESTATE. Mans need fuel and water to conquest space. The moon can supply this for us. We can use a nearly continuous stream of cargo containers coming up from the Moon to both build and fuel space exploration and mans expansion into space.

Different materials could be shot to different receiving areas. Water to one, air to another and steel to the blast furnace.

4. Solar Furnace We can use sunlight and a BUNCH of magnets to create a blast furnace capable of smelting lunar material into stainless steel. This is basically a circular race rack for pieces of Iron, concentrated sunlight would be focused on portions of the stream to gradually heat the iron or steel stream until it is at the desired temperature.

It could also be a stationary glob of white hot metal. A glob of 200,00 tons of molten white hot steel.

We can use strong secondary magnetic fields to magneticly siphon the desired amount of molten iron or steel from the 2,000 degree flow. It would never be allowed to cool. Using magnetic conduits this flow could be directed into a nozzle that puts a high positive charge on the molten material and gently directs the hot metal towards a form that is grounded. This steel could also be directed into very large vacuum casting. The vacuum of space makes it easy to cast things.

5. Steel Banding We can make continuous steams of (1 inch to over a foot wide) steel banding and use this to make VERY LARGE STEEL SPHERES. Steel banding is perfect for creating a woven and meshed surface that is very strong and flexible. This can be used to create much bigger SPHERES.

We can use multiple bandobot's to weave a steel structure. Even passageways and air ducting could be formed from woven steel banding.

5. Relaying Supplies to Other areas From L1 or L2 we can RELAY materials to habitats in space or to Earth. As it is fairly easy to use magnetic guns to shoot material from L1 & L2 to other places in space such as the Earth's L2 (in the shadow of the Earth, away from the Sun). This is another great place for a Colony, I dupe it Soleven. We can make magnetic guns capable to firing objects at 2 miles a second.

Steam a Propellent in Space

Water heated several hundred degrees and kept under several hundred pounds of pressure makes a wonderful propellent. Especially if you are in space already and you don't need to accelerate to escape velocity. When you are not in a hurry steam is a great propellent, it works really well and is safe to use. Maybe a waste in the emptiness of space but a huge amount of thrust is gotten out of just 100 liters of water. But you would not use much in most cases. If you are in a hurry you buy liquid oxygen and hydrogen.

Compare steam to mixing liquid oxygen and hydrogen together in a thruster chamber - now that is dangerous.

Large woven steel conduits would carry air/water to/from the greenhouses. Warm air from the solar heaters and hydroponic gardens would be carried to the living areas, both water and air would flow in these air vents. These same conduits also double as a transportation system. If you were headed for the green houses, you would be riding in a flow of cold air/water returning to the gardens. If you are going home from the green house, you would be riding a warm air current in a conduit leaving the gardens.

In zero gravity it is the kind of place you could keep a fish alive in a bubble of water that floats in your living room. In fact fish and man would coexist in this situation as the fish would likely learn to jump from bubble to bubble. With no gravity it is hard to fill a water tank. So difficult that 50 % or more would likely be considered full. Surface tension would gradually cause most water droplets to gradually come together to become a simple big glob of water. It would be cool to keep a big blob of water and air in a sphere. You just dive into it. Would be fun to use fins and float around in complete leisure.

Habitats on the Moon will need to be underground, as it makes no sense to have anything exposed to the huge diurnal temperature changes of the lunar surface. Plus the threat of meteor impact is much higher without an atmosphere. Even a grain of sand would spell your end. We could make concrete domes. On the Earth some of these big domes are up to 100 meters across. On the Moon with 1/6 the gravity of Earth, domes 1/4 a mile across might very well be possible.

I say **DIG DOWN.**

I would feel better with a dome made of steel banding and covered in concrete. This would be covered by many feet of lunar rock and soil. If it is going to get holed – it will be made to self seal.

Exactly how much American Blood is democracy in Iraq or
Afghanistan worth?

NOT ONE DROP

No more empires
bring the troops home

ALL of them,
from the east and the west

10

SHIPYARD AT SOLEVEN

I propose the establishment of a solar magnetic furnace and shipyard at the Moons L2

If we can make spheres, we can make space vessels and habitats; these will make backers of the project extremely wealthy. L2 will a make a great location for a shipyard, because its so easy to get raw material from the surface of the moon up into a stable orbit in L2. Multiple magnets arranged in circles will move pellets of iron in a continuous circle, this will be heated with solar rays into molten steel.

We can use banding robots or wiring robots that are able to create an object. They each have a perfect computer view of what the final product looks like and they each work together to make it happen. Dozens of robots working together making a sphere or shape.

Imagine a dozen arms with electromagnets and spot welders connected to a unit with a big spool of steel on its back. It could be a big spool of steel wire or steel banding. Steel banding is heavier but far more strong.

This will be used to create lunar steel banding. We can use banding to create huge steel armored spheres in space. Start with just one 30 foot band, Then keep adding bands, just bend the steel and offset a single degree per every hoop you add. Every time you cross another band you spot weld them together. You can use one continuous piece of steel to make the entire sphere. Then you shift 90 degrees and you put another layer of steel over this first one. For a hospital ship put on a third layer.

You use spot welding to connect the bands together anytime they cross another one.

Am I the only one that hates the idea of launching premade structures into space? It MAKES NO SENSE

We can build ships in space, not on the planet. We DON'T have to WAIT for Star Trek. We don't know how to launch the Enterprise off the Kansas plains like they do in the movie. BUT WE CAN launch fairly simple machines into space capable of making large stainless steel spheres. That can be made into Habitats, Spaceships, and Greenhouses.

We don't need a moon base to launch material into space, we can launch from Earth. One machine can do most of the work to make the outer sphere. We use 1 inch to 24 inch wide stainless steel bands. The resulting steel spheres are much like paper mache, but far stronger. At least two layers are used and every time two bands cross they are spot welded. These are spot welded every inch or two.

Steel Required to make BIG spheres

Width of Steel Bands used for making in Inches	Number of feet	Triple layered habitats take three times the steel in feet	Sphere Circumference in Feet	Sphere Diameter in Feet
2	5400	16200	30	9
4	10800	32400	60	19
6	32400	97200	180	57
12	64800	194400	360	115

Surfacing of the Spheres

Inner Surface may be Spray on multiple Carbon fiber layers, epoxied on. Multiple layers of 3 inch long fibers of steel wool is sprayed onto epoxied surface

Re-bar may be used and concrete, options other than Steel Spheres. Steel wool and Carbon fiber balloon the inside of a carbon fiber balloon can be sprayed with layers of 2 inch pieces of steel wool and epoxy

Iron or steel re-bar can be also be used instead of steel banding, this would create a sphere that one can coat with concrete, fiberglass, or carbon fibers

Smaller ceramic spheres can be cast directly out of set, then cast in huge solar ovens

Glass spheres can be blown

Iron spheres can be blown in the same manner as glass

Iron Spheres may be directly cast

Wire Spheres may be spun

I prefer stainless steel bands, these are just super strong.

A steel wool and carbon fiber balloon is something we can make now.

An alternative to steel or iron is ceramic clay spheres. These can be cast in much the same manner as their earthly counterparts. Pour the set into a mold, wait awhile and pour the liquid part out. Take it out of the mold, dry it and prepare for firing (cut all the holes you need for access). Use a solar oven to cook it at the required temperature. This would create an extremely hard casing that can be used to transport supplies such as food, water, or gases. Their biggest drawback is how they fail, get a crack and it might self destruct (unless a lot of carbon fiber or metal was used to reinforce the structure).

A Wire Sphere is cool to imagine building, imagine four or eight or a thousand wirebots running round and round as they make a sphere or special shape. Every time two wires cross, they get welded together. Each wirebot is carrying a couple hundred pounds of wire, but its weightless and they can't tell. But there is a lot of inertia if you move too fast. This is why everyone moves slowly in space. You just can't move fast.

Unless a bomb was exploded within, a sphere made of interwoven steel bands would not fail in explosive fashion. Its just much too strong. Should it be holed by a fast moving asteroid, it would likely be possible to stop the release of air and to quickly patch it

If we can get a supply of iron from the moon and have a foundry in space to make things out of iron, we are light years above anyone. If we can just make banding we can pretty much use that to make anything that we want to make.

Imagine a structure made of a million spheres, a long spinning tube of huge interconnected spheres. The ends and sides have been covered in steel and sealed. It holds over a square mile of air and water. The outer areas have gravity but the center does not.

11

SPACE SHIPS

Lets make Space Ship's and Eco sphere's in space Out of Steel Banding.

We need steel banding. Preferably long continuous bands. This would come out of a dispenser and go into a bender that is able to put out a long continuous bend on the metal. This results in the steel banding curling up on itself. Like paper mache made of steel instead of paper. Use multiple layers and every time one crosses another – you use a spot welder and join them together.

We make a robot that does all of that ans use steel or iron to create shapes and objects using bands or wire, or a more re-bar like construction, use concrete to seal the steel and to create an air tight boundary.

Set 30, 60, 120, 360, 720 feet sizes as a standard - globes would be easiest, these would be rotated rapidly to shape the steel and give it a clearly marked equator

1. Create a strong equator, loop 3 times and weld them together into a tight loop

2. Start another band, but shifting the wire away from the plane of the first loop a set amount – enough so it shifts 1 degree

3. Go a full circle spot weld it and then do it again 178 more times (a full circle)

4. Spot welding the bands together, every time they cross

5. After you get the first layer of the sphere completed, you shift 90 degrees and lay another one down that crosses the first at a 90 degree angle. Spot weld them together everywhere they cross another band.

6. Seal the interior with multiple layers of steel wool, fiberglass, or carbon fibers, simply spray on epoxy and spray the fibers in it.

7. One might use rebar and concrete, this would be a strong and heavy globe

8. For lessor spheres use carbon fiber and epoxy to seal the fabric mesh of the steel or iron wires or bands

These spheres would be very strong and could be used for any purpose: ships, habitats, and greenhouses are a few examples.

We can weave almost any shape out of steel banding or wire.

We can use dozens of banding robots, these would be self propelled and carry rolls of banding (or wire) and could be programed to create multiple shapes. These would work together to create extremely strong steel mesh shapes.

Almost any shape could be woven out of steel:

- Spheres that can be used as ships

- Domes that could be used on the Moon or Mars

- Semi-flexible insulated passageways could be made that are air tight, these could be used to connect various spheres together

- Woven steel is really critical to developing into space.

Imagine going to work harvesting grown plants, fish, and shrimp in the greenhouses or fish farms. I imagine shrimp, crab and rice would both grow well in weightless or near weightless conditions. You would make your way to the exhaust vents in the main habitat and jump in, allowing the current of air to carry you into the woven steel conduit connecting the spheres, it would carry you to work. As you harvest plants in the day you pack them into round steel containers. You simply place these round steel balls in a different air duct and the gentle current of air carries the product or grain to the habitat. At the end of the day, the warm return conduit carries you effortlessly back home.

Conduits between spheres in space would carry mixed air and water, this could be a very wet environment. Imagine if air and water were mixed at 50/50 in a large (non spinning) sphere, it would aquatic in nature, lights or sky lights would bring in light and plants would flourish. Fish and birds could coexist together and fish would be able to jump from bubble to bubble. We just have to design an ecosystem where the creatures keep themselves fit by trying to evade getting eaten. Shrimp and crab would also do well, the calcium from these creatures shells could enrich our diets.

Waters wonderful, we love to live in and be close to it. Yet split it into hydrogen and oxygen and it makes a great rocket fuel. The water could come from the moon, shot up to L1 and L2 in iron cargo containers.

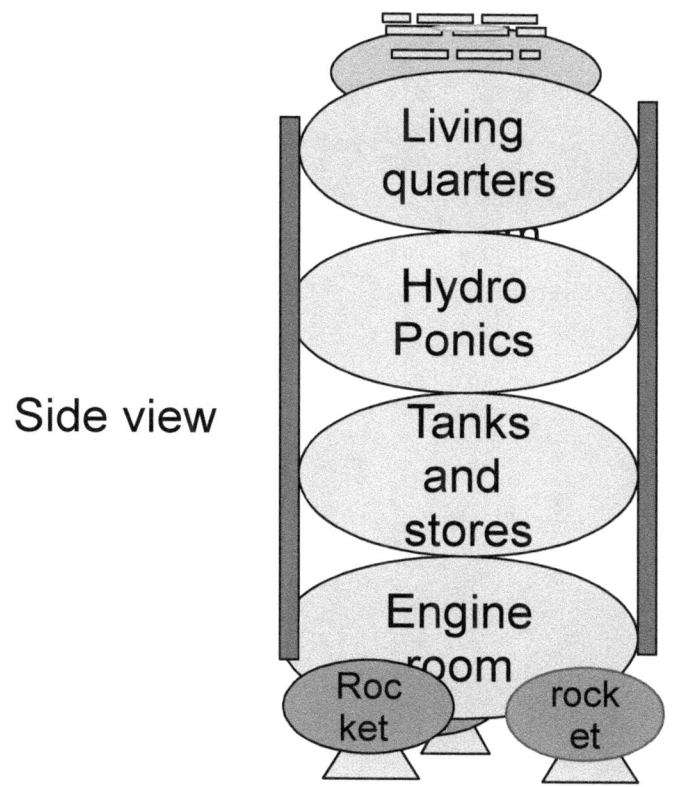

Graphic 1. A SIMPLE SPACE TUG: Spheres give you a multitude of ways to put together a space ship. This is a real simple one.
I would suggest 39 feet as the standard size of for steel spheres, other materials might have different standard sizes.

A standard ship would be:

1. Solar Powered

2. Low maintenance

3. Simple design

4. adio controlled

4.

12

LUNAR ROVERS

Whenever the sun starts shining (2 weeks of day followed by 2 weeks of night), these rovers would start working, endlessly roving the surface of the moon. Using Magnetic pick ups these rovers would reap the iron pellets that cover the surface of the moon. These pellets are the result of high speed iron meteorites striking the moon (maybe 1% of impacts on the moon), these impacts turn bout half the mass in iron meteors into molten metal and these droplets get launched in all directions. The end up as iron pellets on the ground. Should be easy to pick up with a magnet.

The rovers place the iron gathered up in a rover bed – a truck. When the bed is full the rovers would return to a specified location and dump their load. Alternatively the rover would melt and cast this iron into cargo containers.

The rovers would also continually gather and heat solar soil. This would would release water and gases trapped int the soil. These are continually harvested and saved in cargo containers.

A solar oven on the surface of the moon smelts the iron into lunar steel and this would be cast into cargo carrying spheres. These could be filled with steel pellets, oxygen, water, food, lunar tobacco, the list goes on and on. Using a magnetic gun these iron spheres can be shot straight up into L1 or L2.

A radio or light can be the target, and a large net or magnetic field could be used as capturing device. One could even use magnet tipped lines to catch containers as they would be moving very slow.

ON THE MOON

IT'S A PRETTY NEAT TRICK

YOU THROW UP AN IRON SPHERE

AND

IT DON'T COME DOWN

The trick is that you would have to throw it up at just a little higher speed than escape velocity. We can make magnetic guns that shoot faster than the escape velocity of the moon, its around 8,000 miles per hour. The standard sized cargo iron or steel containers would not even touch the barrel of the gun, no wear on the barrel other than stress at the points where the magnets come together.

MICHAEL JOSEPH LITTLE

13

WEAPONS IN THE HIGH GROUND

Molten iron or steel could be shot directly from a magnetic gun without a container.

Easier and faster than using containers. It would also make a fearsome weapon.

It's very hard to argue with someone who is able to accurately throw molten steel or a solid spheres of iron at you at 8,000 – 12,000 miles a hour. Perhaps fearsome enough to think that peace on Earth can be an option worth considering. But if nuclear weapons couldn't do it then this will likely fail also.

Imagine if the Red Chinese, North Korean's or Iranians were able to get a weapon like this in space. They would be able to make a serious impact on the political structure of the world.

They could tell people and governments what to do. If an enemy built a magnetic gun on the moon with an unlimited amount of material to launch They could take the world back into the stone age You just hit anything with a significant infrared heat signature or smokestack It's pretty hard to argue with someone capable of raining fire down onto your head It is the stated political and rebellious goal of many extremest groups To do this very thing

If the world were to be threatened by a rocky asteroid or comet, a magnetic gun on the moon could be used to try and break it up. It could be a world saver. Do we really want the Red Chinese, North Korean's or Iranians get to the moon first?

THE END

so far

MICHAEL JOSEPH LITTLE

ABOUT THE AUTHOR

In January of 1999, I retired from the United States Navy as a Aerographers Mate First Class (a weather forecaster), have since earned a Masters Degree in Business and a Bachelor Of Science Degree in Database Management. After retiring I started a second career in computers and for the last 14 years I have been a computer consultant to a variety of clients and continue doing that on a day to day basis.

I spent 20 years in the Navy, serving during the cold war of the 70,'s 80's and 90's. A curious career path for a liberal and pacifist, due to a combination of patriotism and desire to stay out of trouble, certainly was not for the money. I am a United States Navy Aerographer's Mate (retired now – but that is my rating) and proudly wore my uniform. Was a pleasure to do so and would do it again.

MICHAEL JOSEPH LITTLE

MY CROW

Aerographers Mate 1st Class
Michael Joseph Little
United States Navy,
Retired

MICHAEL JOSEPH LITTLE

In all your journeys

MAY YOU

HAVE FAIR WINDS

AND

FOLLOWING SEAS

MICHAEL JOSEPH LITTLE

Life

As you go about your mundane day

Remember you have won the equivalent of the Universal lottery

Imagine the size of time and space we live in a Universe at least
5 to 10 billion years old

we are in the middle of a vast Universe that stretches outward
5 to 10 billions of light years

Chance

in all this time and space

has selected this very instant for us to be alive

a place where life has taken hold and wildly proliferated

Every single one of us should be thankful to be alive
in such a place of wonder

Mike Little 2011

MICHAEL JOSEPH LITTLE

THE HIGH FRONTIER

MICHAEL JOSEPH LITTLE